HEAT

Rebecca Woodbury, Ph.D., M.Ed.

Gravitas Publications Inc.

HEAT

Illustrations: Janet Moneymaker

Heat
ISBN 978-1-953542-15-1

Published by Gravitas Publications Inc.
Imprint: Real Science-4-Kids
www.gravitaspublications.com
www.realscience4kids.com

Photo credits: Cover & Title Pg: By varts, AdobeStock; Above, By ulga, AdobeStock; P.3. By xamtiw, AdobeStock; P.5. By vladdeep, AdobeStock; P.7. By zmijak, AdobeStock;

What happens when you take an ice cube out of the freezer?

It melts!

What happens when you put a kettle on a hot stove?

What happens when
you light a candle?

Remember that everything is made of atoms.

But did you know?

Atoms are always wiggling!

Review: ATOMS

- **Everything** is made of **atoms**.

- **Atoms** link together to make **molecules.**

- **Atoms** are always wiggling.

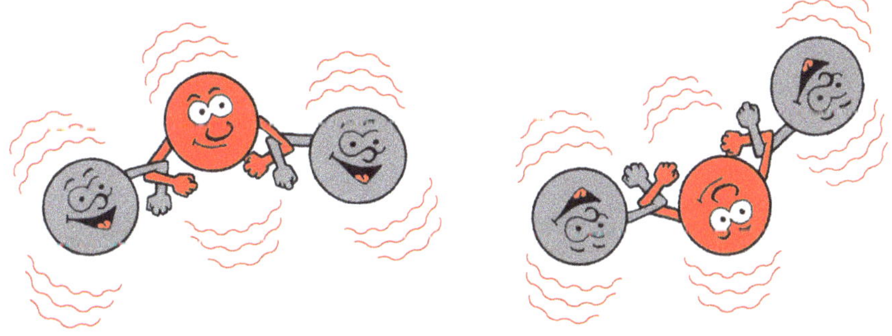

Your hand and a rock are both made of atoms.

If you leave your hand on a warm rock, your hand will start to feel warmer and the rock will start to feel cooler.

This happens because the atoms in the rock that are wiggling faster bump into the atoms in your hand that are wiggling more slowly.

Heat is not a thing, but an action. When two atoms or molecules bump into each other, heat is the action of "wiggles" spreading from one atom or molecule to the other atom or molecule.

This is similar to what happens on a pool table when a ball that is moving fast hits a ball that is moving slowly. The fast ball slows down and the slow ball speeds up.

SMACK!

When we touch a cup full of hot chocolate, we feel heat. The atoms in the cup are wiggling FASTER than the atoms in our hand, so the cup feels HOT.

Cold is what we feel when we put an ice cube in our hand. The atoms in the ice cube are wiggling more SLOWLY than the atoms in our hand, so the ice cube feels COLD.

WE WIGGLE FAST!
YOU FEEL HOT.
WE WIGGLE SLOWLY.
YOU FEEL COLD.

What happens if you put a small piece of ice on your hand?

Can you explain why the ice in your hand melts?

Hint:

- What is it that does the wiggling?
  (atoms in your hand and atoms in the ice)
- What happens when the faster wiggling hand atoms bump into slower wiggling ice atoms?
  (The wiggles from the hand atoms make the ice atoms wiggle faster.)
- What happens when ice atoms wiggle faster?
  (The ice turns to water.)

How to say science words

atom (AA-tum)

heat (HEET)

molecule (MAH-lih-kyool)

wiggle (WIH-guhl)

www.ingramcontent.com/pod-product-compliance
Lightning Source LLC
LaVergne TN
LVHW060634110826
845147LV00014B/909

* 9 7 8 1 9 5 3 5 4 2 1 5 1 *